MONKEYS & APES

A PORTRAIT OF THE ANIMAL WORLD

Paul Sterry

SMITHMARK

This edition published by SMITHMARK Publishers Inc.,
16 East 32nd Street, New York, NY 10016

SMITHMARK books are available for bulk purchase for sales promotion and premium use.
For details, write or call the manager of specal sale,
SMITHMARK Publishers, Inc.,
16 East 32nd Street, New York, NY 10016; (212) 532-6600

This book was designed and produced by
Todtri Publications Limited
P. O. Box 20058
New York, NY 10023-1482
Fax: (212) 279-1241
Printed and bound in Singapore

ISBN 0-8317-0956-1

Author: Paul Sterry

Producer: Robert M. Tod
Book Designer: Mark Weinberg
Photo Editor: Edward Douglas
Editors: Mary Forsell, Joanna Wissinger, Don Kennison
Production Co-ordinator: Heather Weigel
DTP Associates: Jackie Skroczky, Adam Yellin
Typesetting: Command-O, NYC

PHOTO CREDITS

INTRODUCTION

Japanese macaques are one of the few species of apes and monkeys to live away from the tropics. They are found in Japan, and their thick fur helps them survive the harsh winters.

Apes and monkeys are our closest living relatives and it is impossible not to find them fascinating. In their behaviour we see a reflection of our own. While it is inappropriate to anthropomorphise—that is, to interpret their behaviour purely in human terms—we do, nonetheless, have a lot in common with them. A study of apes and monkeys can tell us, on occasion, more than we might care to know about ourselves.

Apes and monkeys range in size from the tiny pygmy marmoset to the huge gorilla. Despite this considerable diversity in physical appearance, apes and monkeys show surprisingly little

variation in their overall anatomical plan. What does vary enormously, however, is their behaviour and the way in which they adapt to changes in their environment. Intelligence and the ability to learn are the keys to their survival, and in that respect we owe our common ancestors a debt of gratitude. Without the ability to develop and adapt these characteristics, Homo sapiens would not have advanced in the way the species has.

In this book, apes and monkeys are discussed initially in the context of their wider classification as primates. Their adaptations to life will be explored as well as their fascinating and complex social behaviour. Lastly, although most people today find apes and monkeys intriguing, their relationship to humankind has not always been mutually beneficial. Some species are hunted for food by indigenous peoples and habitat loss is a major threat around the world. Several species, ranging from squirrel monkeys to chimpanzees and orangutans, continue to be exploited by the pet trade. All these threats are discussed, together with positive conservation measures and opportunities to observe these fascinating mammals in the wild.

Chimpanzees live in fairly loose community groups comprising females, juveniles, and a smaller proportion of males. This group from West Africa is viewing intruders into their territory with suspicion.

This young orangutan is moving through the trees with complete ease and confidence, using both hands and feet to assist its progress. In bright light the orange-red hair is conspicuous but in dappled conditions the animals are surprisingly difficult to see.

A PANORAMA OF PRIMATES

Biologists concerned with the classification of the animal kingdom place apes and monkeys in a taxonomic group called primates. This particular category is called an 'order' and is the one which humans also belong to. As well as the more highly developed apes and monkeys, the primates include more primitive representatives, known as prosimians—the so-called lower primates—which show few similarities to man.

The primates order contains around 180 species and is broadly divided into two major divisions: the prosimians, or lower primates, and the monkeys and apes, referred to as higher primates. The lower primates include lemurs, the aye-aye, bush babies, lorises, and tarsiers. The higher primates include all the other groups, from both the New World and Old World. Included are howler monkeys, capuchins, marmosets, baboons, and langurs; the lesser apes (the siamang and gibbons), the great apes (gorillas, chimpanzees, and orang-utans), and, lastly, humankind.

Almost all primate species are found in the tropical regions of the world. Rain forests display the greatest diversity and abundance. The reason for this global distribution is linked with food and feeding. Although omnivorous, primates feed to a great extent on fruits, nuts, young leaves and insects: equatorial and tropical regions offer these food supplies more or less year-round while they are seasonal in temperate regions. Exceptions to this generalisation include humans and two species of monkey—the Japanese macaque and the Barbary ape (the latter, despite its name, is a macaque, not an ape).

The lifestyles of apes and monkeys are often described by the precise niche that they

Swamp forests in central and eastern Africa are home to the strange but elegant de Brazza's monkey. The most striking aspect of this species is the ginger brow and the long, white beard.

L'Hoest's monkeys live in montane forests of central Africa, their range overlapping with that of the mountain gorilla. Group sizes are small with a single dominant male in command.

occupy within their environment. None lives underground, and so they are characterized as either terrestrial—living on the ground—or arboreal—living in the trees. Some species can, quite happily, make the transition between the two zones, while others may be confined to one or the other through adaptations in anatomy and behaviour.

Throughout the evolution of apes and monkeys, there has been a trend toward increasing brain size, both in relative terms and in absolute terms. With this has come increasing intelligence and an ability to learn. Apes and monkeys are social animals and their social structure and behaviour patterns have also become complex and varied, especially in advanced species such as the great apes. The development and maturity of young apes and monkeys has also become longer as the group has evolved and, as a consequence, their lifespans have increased.

Male western-lowland gorillas characteristically show the hair on the head slightly chestnut or ginger. This picture clearly shows the peaked head so typical of mature male gorillas.

Following page: Large male mountain gorillas are often referred to as 'silverbacks' for obvious reasons. This silverback will be the only fully mature male in his group, the focal point of their behaviour and daily activity.

From Lemurs to Bush Babies

Perhaps the best-known members of the lower primates are the lemurs, unusual-looking primates with dog-like faces. Together with the closely related dwarf and mouse lemurs, the sifakas, the indri, and the aye-aye, lemurs occur only on the island of Madagascar, off the east coast of southern Africa. How they arrived here in the absence of any other primate species is something of a mystery. However, arrive they did and over the last thirty million years or so, they evolved into an array of adaptive forms of which, sadly, only ten remain today.

Ring-tailed lemurs are familiar to many people from captivity but are also widespread in deciduous forests on Madagascar. Like other lemurs, they are mostly arboreal, but groups, typically of five to thirty animals, often move across the ground holding their striped tails aloft. Like other true lemurs, ring-tailed lemurs are essentially vegetarians, feeding on leaves, fruit, and flowers.

There is surprisingly little geographical or habitat overlap between the different lemur species. The most common species—the red-fronted lemur—is widely distributed in west coast forests while the ruffed lemur is found only in east coast rain forests and the brown

The long, banded tail of the ring-tailed lemur is its most distinctive feature. The animal is diurnal and the tail is used to signal mood and intention to other members of the family group.

Taxonomists disagree as to whether the red ruffed lemur is a race of the ruffed lemur or a separate species. Whichever is the case, it is very rare, found only in small areas in the northern part of eastern Madagascar's coastal rain forest.

Verreaux's sifaka is one of the largest of all modern-day lemurs. It lives in groups of five to ten animals and is active during the hours of daylight. A sifaka's diet includes bark, leaves, and fruit.

Black-and-white ruffed lemurs occur in the remaining stretches of rain forest on the east coast of Mada-gascar. Pairs often seem to remain together for life and females usually give birth to twins.

lemur occurs mainly on the northwest coast. Although diet and distribution may differ, all share the characteristically long tail (longer than the body), have arms shorter than the legs, and an elongated muzzle.

Madagascar also has four species of dwarf lemurs and three species of mouse lemurs, all of which, as their names suggest, are small and weigh less than 500 grammes (17.5 ounces). They are nocturnal, almost exclusively arboreal, and have an omnivorous diet,

taking small animals such as insects as well as fruit and leaves. The remaining species of lemur-like primates from Madagascar are the sifakas and the indri, which are arboreal and leap from tree to tree, and the aye-aye. The latter is a bizarre, nocturnal primate with huge eyes and koala-like ears.

The remaining lower primates are found in Africa and Asia and all are distinctive—not to say bizarre—to look at. These include the bush babies and pottos from Africa, the

Despite their common name, it is only the male black lemur that is this colour; females are rich brown. This species lives in groups of five to ten animals in the forests of northwestern Madagascar.

lorises from Asia, and the tarsiers from southeast Asia. The creatures are nocturnal, with huge, forward-facing eyes, and arboreal, having hands and feet with powerful grips.

Movement has been refined to two extremes in these prosimians. Bush babies are renowned for their leaping abilities among the tree tops, even after dark. Tarsiers too are skilled at jumping between branches in the rain-forest canopy. They have dispro-portionately long hind legs and long and slender fingers and toes, the tips of the digits bearing suckers to help grip. Both groups of animals are carnivorous, insects and lizards being favourite foods. Excellent hearing augments their powerful night vision. At the other extreme, the two species of lorises move at an incredibly slow pace. Although they will feed on animal matter, a considerable amount of plant material is included in their diet.

The lesser bush baby is found in a wide range of forest habitats across much of sub-Saharan central Africa. Its large eyes are used both for judging distance while climbing and for locating insect prey, fruits, and flowers.

Lesser bush babies spend the hours of daylight hidden in holes in trees, where they often build themselves a nest of dry leaves. It is not until complete darkness has fallen that they venture out.

Monkeys, Old World and New

Monkeys are classified into 133 species and are distributed throughout the tropical regions of Central and South America, Africa, and Asia. African and Asian monkeys have characteristics in common such as close-together nostrils and the presence of toughened callosities on the buttocks for sitting; they are known collectively as Old World monkeys. Monkeys from the Americas are called New World monkeys. The development of a prehensile tail in some species and their widely spaced nostrils set them apart from their relatives in the Old World.

Despite their differences, monkeys from around the world share similarities which set them apart from their more primitive prosimian relatives. In particular, there is a fairly dramatic increase in brain size, both in relative and absolute terms. An increasing dependence on the sense of sight, compared to smell, is reflected in the size of area of brain concerned with this sense, as well as in the positioning of the eyes.

New World monkeys are classified into two main families, the marmosets and tamarins in one, and the capuchin-like monkeys—namely capuchins, titis, sakis, uakaris, as well as howler, woolly, spider, squirrel, and night monkeys—in the other. In general, marmosets and tamarins are small, arboreal creatures that live in small family groups. Many of

Golden lion tamarins are severely endangered and found only in a few pockets of coastal rain forest near Rio de Janiero in Brazil. Their name is derived from the beautiful mane of golden hair.

The Amazon rain forest is home to the red-chested, moustached tamarin, sometimes simply known as the moustached tamarin. Like other tamarins, it lives in small family groups that forage in the tree canopy.

the capuchin-like monkeys are much bigger, and some species live in comparatively large troops.

There are twenty-one species of marmosets and tamarins, all of which are found in northern South America, most living within the Amazon basin or Brazil's remaining coastal rain forests. They are characterised by their long tails, fluffy coats, and the bizarre head tufts, manes, and ruffs in the various species. With the exception of their big toes, marmosets and tamarins have claws rather than nails at the end of each digit. As a general rule, females give birth to twins, pairs are monogamous, and both parents care for the young.

Named for its extraordinary head of hair in a beautiful golden-blond colour, the golden lion tamarin is perhaps the best-known sub-species of lion tamarin. Because of habitat loss and exploitation by the pet trade, it is very rare and only found in small areas of primary forest to the north of Rio de Janeiro. Like most other tamarins, the golden lion tamarin has a varied diet including small animals, such as insects and lizards, nectar, gum, and fruits. The tiny pygmy marmoset, on the other hand, feeds more exclusively on sap and gum, but will take small invertebrates if easily captured. It weighs around 130

The silvery-brown bare-faced tamarin or white-handed tamarin is a severely endangered species from Colombia. The cheek hairs are long and spread upward, giving the impression of an unusually broad face.

The saddle-back tamarin is sometimes also known as the white-lipped marmoset because of its white muzzle. It occurs in the upper Amazon and the back markings vary according to the animal's site of origin.

grammes (about 4.5 ounces) and is the smallest monkey in the world.

The capuchin-like monkeys vary in appearance. Opinions differ as to their precise classification, but a consensus of opinion suggests that there are thirty species. Most have rather flattened faces, with the nostrils widely spaced. They live mainly arboreal lives, often in fairly large groups, and some have developed prehensile tails to assist their passage through the branches.

The squirrel monkey is one of the better known of the capuchin-like monkeys, partly because it is, regrettably, popular in the pet trade. A range of subspecies occurs from Costa Rica south to Brazil, living in forest types ranging from inundation to highland, and all move by leaping through the branches. Squirrel monkeys live in groups of thirty or more and are omnivorous, feeding mainly on fruit and insects.

The dusky titi is representative of the titis as a whole. It prefers inundation forest and feeds primarily on leaves and fruit. Titis advertise their presence within a territory by loud calls given at dawn, and this habit is shared by that most vocal group of monkeys, the howlers. At close range, the sound is almost deafening and carries for several miles in the still air of the rain forest.

Confined to a region of South America north of the Amazon River and east of the Rio Negro, the white-faced saki is a monkey of the lowland forest. The white face is striking and more pronounced in males than in females.

As its name suggests, Humboldt's woolly monkey has dense fur all over its body and tail. This is surprising, since the monkey lives in the hot and humid environment of the Amazon rain forest.

19

The prehensile tail of this spider monkey must be strong enough to carry the weight not only of its own body but also that of its baby. While the monkey feeds on leaves, the tail alone may have to bear the full weight of the animal.

Looking more like an extraordinary breed of dog, the cotton-top tamarin lives in small family groups in the forests of northwest Colombia. It feeds mainly on leaves, fruits, and flowers.

The calls demonstrate to rivals the group's control over a productive tree. Perhaps the most bizarre calls of all come from the night monkey, a nocturnal resident of the Amazon rain forest. Males that are searching for a mate utter eerie hoots, usually on moonlit nights.

Various spider monkeys can also be found in the forests of Central and South America, all with prehensile tails. This adds freedom when moving through the branches in search of fruits and leaves and can leave both hands free when feeding. The woolly monkeys are represented by two species, Humboldt's woolly monkey being widespread but locally distributed, living in groups of up to twenty animals. As their name suggests, they have thick fur.

Surely the oddest member of this group of monkeys is the red uakari, which lives in restricted pockets of inundation forest in the central Amazon. Male red uakaris have

Weighing less than 150 grammes (5.25 ounces) and with a head and tail length of only 40 centimetres (15.75 inches), the pygmy marmoset is the smallest monkey. Its diet includes sap, which it gathers by gnawing through the bark of the tree with its specially adapted teeth.

Looking like it has a bad case of sunburn, the male red uakari has a bright red face. Opinions differ as to the reason for this colour but the degree of redness may indicate the monkey's state of health to a potential mate: the redder the face, the healthier the individual.

Coat colour in the red uakari is very pale in some populations, which may explain the alternative name of white uakari. In the wild, the monkeys live in groups of twenty to thirty animals and are found in low-lying inundation forests.

Following page:
Denizens of humid forests from sea level to 2,000 metres (6,600 feet), white-faced capuchins occur from Belize and Costa Rica in Central America south to Colombia. They live in large, noisy groups of ten to twenty animals.

23

bright red, hairless faces; they are known colloquially as the English monkey, perhaps because of their resemblance to a sunburnt tourist.

The Old World monkeys are represented by eighty-two species, which include the leaf and proboscis monkeys and macaques from Asia; from Africa come the baboons, mangabeys, guenons, colobus, and patas monkeys. Most are typically monkey-like in appearance and many spend a considerable amount of time on the ground. The mandrill and the drill, both species of baboons, are atypical with enlarged heads, elaborately marked faces, and reduced tails.

Macaques have a wide distribution throughout southern Asia, mostly with little species overlap. Although feeding primarily on plant matter, macaques are omnivores and skilled foragers. Crab-eating macaques, for example, will catch crabs or almost anything else they can eat in the tidal and riverine forests they inhabit. Not all macaques are lowland or tropical species: the rhesus macaque lives in the foothills of the Himalayas and the Japanese macaque lives in temperate regions of both main islands. Most macaques are Asian, but one, the Barbary ape, occurs in the Atlas Mountains in North Africa as well as on Gibraltar.

Baboons are exclusively African in their distribution. Sometimes split into different

species or subspecies, the savannah baboon is the most widespread. Large troops, often more than fifty strong, can be seen foraging in most areas of grassy savannah in Africa south of the Sahara. The Hamadryas baboon in the highlands of Ethiopia behaves in a similar fashion. Drills and mandrills are both forest species from West Africa that live on the ground.

As a group, the mangabeys are superficially similar to baboons but have well-developed

The Celebes macaque lives in the forests of Sulawesi and is an endangered species. The jutting brow above the eye and the expanded and ridged face give this species a bizarre facial appearance.

Sometimes even a thick layer of fur and access to hot springs is not enough to keep out the cold. The pair of Japanese macaques shown here are huddling together for extra warmth on a freezing winter day.

The rhesus monkey is perhaps best known for its long and long-suffering association with medical research. Despite its popular name, it is a macaque, not a monkey, and occurs in the wild from India northward to Nepal.

The natural range of the Barbary ape, or Barbary macaque as it should be known, is North Africa. Introduced to Gibraltar for sport in the eighteenth century, they remain to this day a feature of this famous rocky outcrop off the southern coast of Spain.

Gelada baboons live in the grasslands of the Ethiopian highlands. They have rather shaggy fur on the head, neck, and shoulder and show bare patches of reddish skin on the chest and neck.

tails. All are forest-dwelling species and have large incisor teeth which they use to crack hard nuts and seeds. The guenons are also a forest-dwelling group of monkeys containing a number of closely related species. Best known of these is the vervet monkey, which occurs as several races throughout its extensive range in sub-Saharan Africa. It is an omnivore, living in large troops in association with savannah woodland.

The final division of monkeys contains the leaf monkeys, the langurs and proboscis monkey and relatives from Asia, and the colobus monkeys from Africa. The proboscis monkey is the most bizarre-looking of the Old World monkeys. Males, considerably larger than females, have long, flattened, pendulous noses, reminiscent of male elephant seals. Proboscis monkeys live in mangrove and riverine forests of Borneo. In common with the closely related snub-nosed monkeys, their diet consists mainly of leaves. Large stomachs assist in the process of bacterial digestion.

Many of the langurs and leaf monkey species are forest dwellers and have little contact with man. The silvered leaf monkey is an exception and can often be seen in

Widespread in southern Africa, Chacma baboons live in large groups. Care of the young is mainly provided by the mother. In some circumstances, however, 'aunts' also offer help.

Few animals can rival the striking appearance of the male mandrill. Relatives of the baboons, these monkeys live in ground-foraging groups in the rain forests of Cameroon and the Congo.

Proboscis monkeys inhabit mangrove and riverine forests in Borneo. Living in such a watery environment, it is not surprising that they are good swimmers, often crossing wide channels to reach new feeding grounds.

The facial skin of female and juvenile mandrills is less colourful than in the males. They are, nevertheless, still very striking animals that can put up a robust defence against most predators except man.

villages and temples and at the roadside. Its range includes Thailand, Borneo, and western Malaysia. The Hanuman langur, too, has a fairly close association with man. Revered by Buddhists in its southern Himalayan homelands, the langurs are often fed by people and subsist to a large extent on agricultural crops.

In Africa, there are four species of black colobus monkeys and five species of red colobuses. All are relatively long-legged and long-tailed and are forest-dwelling species. The best-known species are the Guinea forest red colobus from West Africa and the Guereza or eastern black-and-white colobus, which occurs from western Kenya into central Africa.

The impetus for the name white-epauletted black colobus, and the alternative—black-and-white colobus—is plain. Careful grooming is needed to keep the coat in good condition.

Newborn primates, such as this Hanuman langur, are almost completely defenceless. The youngster will rely on its mother for food, protection, support, and transport for many months to come.

Great Apes, Chimps, and Gibbons

There are thirteen species of apes, including the gibbons, referred to as lesser apes, and the great apes, namely chimpanzees, gorillas and orang-utans. These are our closest living relatives and much can be learnt by studying them. Apes differ from monkeys in that the arms have become important in movement and are longer than the legs and, unlike monkeys, apes do not have tails.

The great apes need very little introduction. Chimpanzees are divided into two species, the common chimpanzee from west and central Africa and the pygmy chimpanzee—only marginally smaller than its relative, despite its name—from Zaire. The orang-utan, with its beautiful red coat, comes from Borneo and northern Sumatra. Lastly, man's closest relative, the gorilla, is divided into three subspecies and comes from central and west Africa. Social

Moloch gibbons come from west Java and their range does not overlap with that of any other species of gibbon. Males utter loud hooting calls to advertise their presence in a territory.

'Old man of the forest' is the literal translation of orang-utan—a name that seems particularly apt when applied to this photograph. These great apes come from Borneo and northern Sumatra.

A pair of Muller's gibbons from northern Borneo views the photographer with misgiving. Their supreme ability to move through the trees, however, gives them confidence in their ability to escape if danger threatens.

The bonobo or pygmy chimpanzee is only marginally smaller than its more common relative, although its body build is less robust. The black face and sideways-projecting hair tufts are characteristic of this species.

A baby bonobo clings to its mother's hand for security. The last remaining stronghold for this species is the forests bordering the river Zaire in the central African country of the same name.

organisation in the great apes is complex and parental care is developed to a high degree. All four ape species are threatened by man, both directly by hunting and also indirectly through habitat loss.

There are nine species of gibbons. All occur in southeast Asian forests both on the mainland and on certain islands, and most are separated from one another geographically. Gibbons are monogamous and the sexes are a similar size. The development of the arms and the swinging method of movement has been taken to extremes in the gibbons. They have an essentially vegetarian diet and advertise their territories with loud calls.

A newborn chimpanzee is helpless but within a few days it is able to cling to its mother. It remains in this dependent state for several months more before starting to walk on its own.

Western lowland gorillas come from the rain forests of the Congo, Cameroon, and neighbouring countries. Unlike the mountain gorilla, the silvery-grey fur on the back of the male extends to the rump and thighs in this subspecies.

LIFE IN THE TREES

With the exception of the gorilla, most apes and monkeys spend a considerable amount of time in trees. The amount varies from species to species and is also influenced by factors such as time of day, behaviour patterns, and weather. Some species, such as baboons, sleep in trees but feed and move mainly on the ground. At the other extreme, gibbons are almost exclusively arboreal and spend virtually their entire lives off the ground.

All apes and monkeys have five digits in both hands and feet. However, their structure, proportion, and appearance varies greatly according to species. Many of the

The Sepilok Reserve in Sabah is one of the best places to see orang-utans in their natural environment. In addition to wild populations, many former pet apes are rehabilitated to life in the forest and taught how to fend for themselves.

White-face capuchins are agile climbers, using the tail to assist in both balance and support. They feed on fruit, nuts, and small prey such as lizards and insects gleaned from among the leaves and branches.

There can be few more impressive sights in the remaining forests of Madagascar than that of a leaping Verreaux's sifaka. The long tail helps in balance as the animals jump from one tall cactus-like tree to another.

A completely arboreal lifestyle has been adopted by members of the gibbon family. This white-handed gibbon shows its immensely long and powerful arms which are used to swing arm-over-arm through the branches.

modifications in the basic plan are concerned with assisting movement, but some of the changes in the hand help with other functions such as feeding and grooming.

Gibbons move through the trees in a rather unconventional manner, by swinging from branch to branch with their forearms. These are much larger than the legs. The four fingers on each hand are elongated and suited to gripping branches and swinging. The thumb is sited well back on the hand so as not to interfere with the grip. Spider monkeys, too, occasionally move about by swinging through branches and show a similar reduced thumb.

Orang-utans are the only truly arboreal apes and, like gibbons, rely heavily on their forearms when moving through the branches. However, to assist their progress they have a big toe that, thumb-like, acts in opposition to the other toes and serves to grip branches. A keen grip is a vital necessity for animals moving through trees often one hundred feet or more above the ground. The spider monkeys of South America have taken the process a stage further and developed a prehensile tail—in effect, a fifth limb. Although they occasionally use this as their sole means of suspension from

a branch, the tail is more usually used in conjunction with one or more of the other limbs.

Rapid movement through the trees requires not only a good, solid grip but also sure-footed confidence in the knowledge that a slip could be fatal. Powerful muscles are obviously important, those in the legs providing the impetus in leaping species, and those in the arms being called upon in swinging species. However, all these physical attributes would be of little use were it not for the excellent vision of apes and monkeys. All have forward-facing eyes and binocular vision and can judge distances with great precision.

Seeing and Hearing

Since their first appearance on the planet, apes and monkeys have gradually refined their senses, placing greater emphasis on vision and less emphasis on smell and hearing. These latter two senses are still of vital importance; it is just that their roles have changed somewhat. The sense of smell is less important in locating food for today's apes and monkeys than it is in assessing its nature or quality, for example. Likewise, hearing is relatively unimportant in locating food and avoiding predators than in

The orang-utan is the only truly arboreal great ape and seldom comes to the ground. The powerful arms are used to swing through the trees, with both their hands and feet gripping the branches.

Central America is home to the black-handed spider monkey. Like its close relatives, this monkey has a prehensile tail to assist movement through the trees. It seldom, if ever, descends to the ground.

other groups of mammals. As an adjunct to vision, it is essential, however, for the successfull functioning of the complex communication that underpins the fabric of social life in apes and monkeys.

The changes in the relative importance of the different senses is reflected in the brains of apes and monkeys. Not only has there been a gradual increase in size with evolution, but the relative areas of brain associated with each of the senses has changed. This is particularly noticeable in the part of the brain associated with vision. The eyes, after all, are merely the receptors; it is the brain which interprets the images and creates stereoscopic vision.

Apes and monkeys are essentially diurnal creatures and are ill-adapted to life after dark. An exception to this rule is the night monkey, or douroucouli, the only truly nocturnal monkey, a denizen of the Amazon rain forest. Night monkeys are able to leap from branch to branch, feed on fruit, nuts, and insects, and carry out all activities normally performed during daylight hours by other monkeys. They are, however, adapted to living in low-light conditions and not zero light conditions. Therefore, night monkeys are most active on moonlit nights.

Lastly, the sense of touch is all-important to apes and monkeys, inextricably associated with the sensations of pressure, pain, and temperature. Although the whole skin surface is sensitive, nerve endings associated with touch are concentrated in the hands and feet. In the hands, the sense of touch allows fine-tuning of processes of handling, gripping, and manipulation by allowing coordination of finger and thumb movements.

Monkey Brains

The secret of the success of apes and monkeys, and also of their human descendants, has to do with their adaptable and complex behaviour rather than their physical attributes. Although there is a considerable difference in the outward appearance of the various species within the group, their bodies conform to the same basic plan and there is little doubt they are all related. Behaviour patterns vary tremendously, however, and the ability to adapt is the hallmark of this fascinating group of animals.

Underlying the extraordinary array of behaviour patterns we observe is a powerful brain which has shown a general trend to increase in size and complexity as monkeys and apes have evolved. This permits intelligent responses to new situations based on previous experiences, and allows animals, and groups of animals, to learn new patterns of behaviour readily, thereby increasing their ability to survive and thrive.

In lower animals, a successful behaviour pattern is refined and passed on to successive generations by the process of natural selection: only the most successful members of a population survive to breed. Changes take place very slowly and there is little opportunity for an individual of a species or its offspring to alter its behaviour and take advantage of new opportunities. Intelligence and the ability to learn offer opportunities to bypass the process of natural selection, because offspring can actually be taught how to respond.

Compared to a newborn antelope, for example, a young monkey or ape of a similar

Study the hands of any primate and you will immediately see similarities with our own. This red ruffed lemur's hand clearly shows four fingers and a thumb; it is used both for walking on the ground and for climbing.

age is poorly developed. After a few days, a young wildebeest has to keep pace with its mother and within a year it will be fending for itself. By comparison, a young chimpanzee stays with its mother for six or seven years. During this time it is cared for and is constantly learning the art of survival and social interaction from the group it lives in.

The intelligence of apes and monkeys is nowhere better demonstrated than in the chimpanzee. Patient observation and study of these animals, especially by the distinguished ethologist Jane Goodall, has revealed an extensive use of tools by these remarkable apes. One fascinating method employed is to fashion twigs or grass stems so that they can be inserted into holes in termite mounds. Soldier termites cling to the stems, which are then withdrawn with the insects still attached. The termites can then quickly be consumed before they have time to bite. Not every chimp uses this technique but offspring often learn the method from their mothers. Special stones are also chosen to assist the breaking of hard-cased nuts and fruits. A flattish basal stone is selected together with an appropriate smashing stone. Both are positioned carefully for maximum effect and are used by successive generations.

Although chimpanzees may appear to live harmonious lives, there is a rougher side to their nature as well. In scenes reminiscent of

the beginning of the movie '2001: A Space Odyssey', rival adult males have been seen throwing branches and stones at one another, trying to intimidate and inflict injury. The same tactics are occasionally employed against potential prey.

What's for Dinner?

Although a few species of apes and monkeys have become rather specialised feeders, most are omnivorous. This is not to say that

Manipulative hands with opposable thumbs enable this vervet monkey to feed dexterously. The skilled use of the hands is one of the keys to the success of apes and monkeys as a group.

The range of the concolor or crested gibbon includes Vietnam, Laos, and southern China. In this photograph, the long, slender limbs can be seen, together with the elongated fingers of the hand.

Monkeys are nothing if not curious and inventive, especially where potential food is concerned. This Chacma baboon has learned how to break ostrich eggs by smashing them on rocks or hard ground.

47

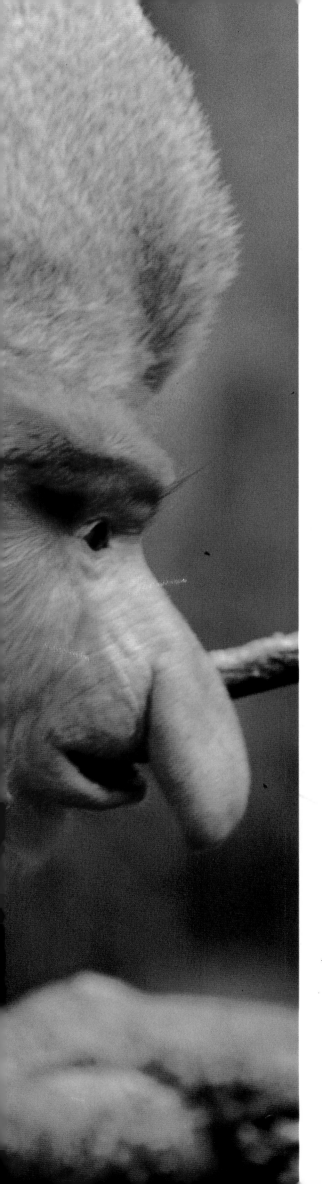

certain types of food do not prevail in the diets of some species, but that most species tend to make the most of whatever food is available. In common with other mammals, the mouth and in particular the teeth are vital for collecting and processing the food for passage and digestion through the gut. Modifications in the basic plan of teeth and gut reflect feeding methods or food types that predominate in a particular species' diet.

The basic dental plan of apes and monkeys is similar to our own. At the front of the mouth there are flat-edged incisor teeth, used for cutting, shearing, and tearing. These are flanked by sharp canine teeth used for piercing and tearing. Along the sides of the mouth are premolar and molar teeth which are flattened and ridged: these grind and macerate food prior to swallowing.

One of the more unusual developments in the teeth can be seen in the marmosets from South America. The lower incisors are angled forward and are used to chisel gouges in the

Keeping a watchful eye open for predators such as eagles, this group of gelada baboons is foraging for roots, nuts, and small prey such as insects. Twenty to forty animals is a usual number for a group of this species.

There can be few stranger sights in the animal kingdom than a proboscis monkey's face. Add to this a swollen and distended belly—needed to digest its leafy diet—and the result is a highly improbable animal!

bark of trees. The animal then laps up sap that is exuded, an important source of sugars and minerals and a supplement to the fruit, leaves, and insects that comprise the rest of the diet.

Leaves form a significant part of the diet in many species of apes and monkeys, the degree of importance varying from species to species. In most, the leaves are a complement to the other food sources such as fruits, seeds, flowers, and invertebrates. As a consequence, the arrangement of their digestive systems is straightforward with comparatively small stomach and small and large intestines of rela-tive lengths comparable with other omnivores.

In gibbons and howler monkeys, however, leaves form a fairly major component of the diet with the result that their large intestines are more developed. This allows for extensive bacterial fermentation of leaf cellulose to take place. Without the action of bacteria, the leaves would provide little nutritional value. In colobus and leaf monkeys, this process has been taken one stage further. These animals feed almost exclusively on leaves and have enlarged forestomachs in which bacterial fer-mentation can take place.

This picture clearly illustrates the size difference between a silverback male mountain gorilla and a mature female. She also lacks the high, peaked head characteristic of the male.

The diet of a gorilla consists almost exclusively of leaves and shoots gleaned from the forest floor. A large and distended stomach helps extract nutri-tional value from this hard-to-digest food source.

SOCIAL BEHAVIOUR

All monkeys and apes live in groups of one sort or another. This social arrangement offers both advantages and disadvantages to its members, but clearly the former outweigh the latter or otherwise the system would not exist. On the plus side, living in a group means that there are more eyes watching out for predators and, when detected, these can be repelled more successfully. Males stand a better chance of mating at a time when the female is capable of conceiving and offspring can receive greater care and attention with group help. On the negative side, there will obviously be increased competition for food within the group.

Group Dynamics

The structure and the composition of groups varies greatly throughout the 146 species of monkeys and apes. Some live in family groups which are, broadly speaking, similar to our own, in that they comprise two parents and their offspring. Other species live in groups where several females and their offspring coexist under the dominance of a single male. Lastly, the third type of group sees several females and their offspring loosely coexisting with several males of different ages.

Generally speaking, gibbons live in family groups. The two adults usually have at least one or two young living with them, these remaining until they are old enough to breed themselves. Because the previous year's young may also live within the group, the size can vary from three to six animals. South American marmosets also live in family groups but these also include offspring that

Grooming is an important part of daily life among most primates, including these olive baboons. Not only are external parasites removed but, more important, the activity helps bond members of the group.

Like other gibbons, the siamang lives in small family groups comprising a male and female, and between one and three offspring. The juveniles normally remain in the group until the female gives birth again.

A year-round water supply is an essential feature of the home range of olive baboons. The whole group will visit the site on average twice a day to drink; at this time they are extremely vulnerable to predators.

As with the common chimpanzee, the bonobo lives in large but loose community groups. They move across the ground on all fours, using their knuckles instead of spread hands, in the manner of gorillas.

Oblivious to the freezing winter temperatures, life for these Japanese macaques goes on as normal. Here a mother grooms her youngster against a backdrop of frozen ice and snow.

Following page: A captivating scene as Japanese macaques lounge in the sauna-like hot springs of their native land. The layer of snow on their heads indicates the severity of the weather in the winter months.

have reached breeding age. While in the family group, however, they do not breed, this privilege being reserved for their parents.

Harem groups—so-called because a single male and several females live together—can be observed in gelada and hamadryas baboons. Inevitably, males are born into the group but as they approach maturity they are ejected by the dominant male and may form their groups of dispossessed males. Gorillas, too, operate within this system; the group is dominated by a huge silverback male.

Multi-male systems can be seen in olive baboons that live in groups of fifty to one hundred animals. Only powerful and assertive males are able to live within the group. Weaker intruding males, as well as their own juvenile male offspring, are driven out by the leading males. Within the group's males themselves there is a hierarchy, with one male dominating the others and, as a consequence, the group as a whole.

Social contact and the essential removal of ectoparasites are important functions of grooming. This behaviour, common among primates, is performed with great relish among members of the same gorilla group.

Midday is a time of rest for gorilla groups living in the wild. This gives the animals plenty of time to start digesting food and also to engage in social contact behaviour such as grooming.

Defending Their Borders

Their environment is all-important to monkey and ape groups. In order to survive, they need a year-round supply of food and water, as well as security from predators by day and, more important in most species, by night. As such, apes and monkeys invariably operate within fairly well-defined areas that provide all their needs. The animals themselves clearly have a good idea of the boundaries and extent of these areas and we, as outside observers, categorize them as either home ranges or territories.

The term 'home range' denotes the area of land used by the group throughout the whole year. On a day-to-day basis, only a small fraction of this will be visited and a troop of baboons, for example, might operate in a completely different area from one month to another. The home range is not defended as such and may overlap with the home ranges of other groups. This is not to say, however, that there will not be conflict if two groups do meet by chance. A territory, on the other hand, is defended. Boundaries are usually well-defined and advertised and there will be very little overlap with the territories of neighbouring groups.

Home range and territory sizes vary greatly from species to species. In general, those in lush habitats such as rain forests are smaller than those in arid regions. Foraging species tend to occupy larger areas than species where leaves form a major part of the diet.

Different species of monkeys and apes can sometimes occupy the same habitat. There is seldom conflict unless they are competing for the same food. They mostly avoid competition even if it is not immediately obvious that they are doing so. This is best demonstrated in tropical rain forests where different species may not only take different foods but may also feed in completely different layers within the forest.

Talking to Each Other

The ability of monkeys and apes to communicate is what underpins the social structure of their families and groups. Communication operates at a variety of different levels from the subtle to the overt, performing many different functions. The role of some of these is immediately obvious while others may take considerable time to discern and unravel.

Although smell is used to a limited extent as a means of communication, visual and vocal signals are far more important means of conveying information among monkeys and apes. For example, sound may be used to advertise the presence of one group of monkeys to another group of the same species in the vicinity. On the other hand, facial expression and posture reflect an individual's mood and can indicate its likely response in a given situation.

Sound is a useful means of communication, especially when monkeys and apes cannot see each other well, as is often the case with forest-dwelling species. Loud calls can be used to indicate the presence of a predator, but in some species they are used for communication either within groups or between them.

Grey-cheeked mangabeys from central Africa utter a loud call which is often rendered as 'whoop-gobble'. When given by a dominant male, it serves to attract and round up members of the group that may have become dispersed. Gibbons, which are territorial and live in family groups, sing loudly, the two partners in a group often dueting. The sound serves

Olive baboons live in large groups of up to one hundred animals comprising females, juveniles, and a core of dominant males. They often forage on the ground, with 'lookout' animals keeping an eye open for danger.

Facial expression in apes and monkeys often gives a clear indication of the mood and behavioural state of the individual. Chimpanzees have been particularly well studied by scientists such as Jane Goodall.

both to demonstrate their presence within a territory but also as a means of maintaining the pair bond. The other well-known group of vocal primates are the howler monkeys of South America. The dawn howling carries for up to a kilometre (.62 miles) and serves to advertise a group's location within its home range to neighbouring groups.

Being primates ourselves, it should come as no surprise that facial expression in monkeys and apes is a revealing mirror of an individual's mood and intentions, at least to other members of the same species. A relaxed face, with the mouth wide open and the upper teeth concealed, indicates a desire to play. A yawn, on the other hand, often indicates nervousness and is often seen in subordinate male baboons in the presence of the dominant male. A wide grin revealing all the teeth generally shows fear while a pouting expression may mean submission. Used in conjunction with gestures and posture, the use of these expressions usually avoids direct conflict between individuals.

Who's the Boss

At first glance monkey and ape groups may appear somewhat disorganised, but watch them for any length of time and it soon becomes apparent that they are well structured. In almost all species, males and females are organised into a hierarchy of dominance, the sexes separate from one another. This affects almost all aspects of their lives, including the ability to mate,

The yawn of this proboscis monkey only serves to exaggerate its extraordinary facial appearance. The flattened, pendulous nose, which is larger in males than females, is used in display.

In primates, yawning is not only a sign of tiredness but often an indication of anxiety. This male olive baboon, photographed in Kenya, is nervous about the approach of a dominant rival male.

The range of the patas monkey extends across central Africa from Kenya in the east to Senegal in the west. The whisker and hair tufts on the head help accentuate facial expressions used in communication.

access to food, and the safest sleeping areas. Generally speaking, males are dominant over females although in many areas of their lives there is relatively little overlap.

Within monkey and ape species that live in multi-male groups, such as the olive baboon, there is a single dominant male. His position, and the retention of it, depend to a great extent upon his strength and confidence, both of which are constantly being tested and challenged by subordinates. His position as dominant male does not last forever: sooner or later, perhaps after the dominant male has been weakened by constantly having to assert himself, a more powerful male succeeds him. Since male offspring born into a group leave it on maturity, his successor is unlikely to be related to him.

With female baboons, the dominance hierarchy often seems to follow a matriarchal line. It would appear that a young female's position within the hierarchy is influenced by the relative dominance of her mother: dominant mothers raise dominant daughters.

The existence of dominance hierarchies within monkey and ape societies does not mean that life is necessarily a constant battle of assertion. Species have evolved lots of different ways of smoothing over differences and demonstrating submissiveness without having to resort to putting it to the test physically. Facial expressions and presentation of the rear end can demonstrate submission, while grooming is a widely used form of appeasement.

Gibbons are among the most vocal of all apes. Male siamangs utter loud hoots accompanied by barking calls from the female. These serve to advertise their presence in a territory and to reinforce the pair-bond between them.

Play is an essential part of everyday life for young Japanese macaques, as indeed it is in all other primates. It helps them learn skills vital for survival and establishes a dominance hierarchy among peers.

The combination of facial expression and vocalisation can be very expressive in species such as this Sykes monkey from Kenya. Generally a single adult male dominates a group of twenty to thirty animals.

Mothers and Babies

The perpetuation of the species is the single most important aspect of an animal's life, those of monkeys and apes included. Armed with this knowledge, it is not difficult to appreciate how much of primate behaviour affects and is affected by the need to reproduce.

In some animals, care for the young lasts for only a short while after birth. In monkeys and apes, however, infants are born in a completely defenceless state and have to be cared for over an extended period which may last from months to years. Even when the youngster is able to fend for itself somewhat, it remains in close contact with its mother. It is not until puberty—the onset of sexual maturity—that close ties are severed. During the intervening two to ten years, depending on the species, the young primate is not only protected by its mother but learns from her as well.

Female gorillas usually start to breed at around ten years of age although they may mature a few years earlier. A newborn youngster may weigh as little as 2 kilogrammes (about 4.4 pounds) at birth but rapidly puts on weight, thanks to its mother's milk.

This female white-epauletted black colobus is cradling her newborn youngster. This monkey is distinguished by having a mantle of long and elegant white hairs contrasting with the otherwise black fur.

Newborn gorillas are defenceless and unable even to crawl at first. After a couple of months they develop this skill but it takes at least eight months before they start to walk.

A young chimpanzee begins to explore the world around it and develop the skills necessary for survival. It will also imitate the actions of other, older chimps in order to learn these skills.

By the standards of other primate species, the young gorilla is a slow developer. It may take three years before it is fully weaned and it may stay in fairly close contact with its mother for several more years.

This female ring-tailed lemur is carrying her six-week-old twins on her back. Parental care for the young for extended periods is characteristic in all primate species.

Young gorillas are breast-fed with their mother's milk for the first couple of years of life. During its early months, the female shows an almost obsessive devotion to the youngster, which would otherwise be extremely vulnerable to predators.

Male monkeys and apes are usually willing to mate at any time. Females, on the other hand, have a reproductive cycle and are generally only willing to mate, or are only attractive to males, around the time when they can conceive. At this time—called oestrus—clear physical and social signals are given by the females of most species that they wish to mate, at the optimum time for conception. Gestation in monkeys and apes can last from as little as 145 days in marmosets to 264 days in the orang-utan, a period that is not far short of the period of pregnancy in humans.

Because of the dangers associated with giving birth, and in particular the mother's vulnerability to predators, most smaller species of monkeys and apes give birth at night. A single offspring is usual in most species although marmosets generally give birth to twins. With the exception of the great apes, most primate young show an innate ability, from an early age, to cling to their mothers—a trait seemingly essential in arboreal species. During its infancy, the young monkey or ape feeds on milk from its mother and is gradually introduced to solid food after a period of several months.

During infancy, the mother monkey or ape not only provides food for her young, but also protection and the opportunity to observe and learn behavioural and social skills. As its development progresses, this role is also filled by other members of the group in which it lives. These guardians or teachers are usually older brothers or sisters, the mother being generally reluctant to entrust the care of her young to individuals not related to her. Care for the young comes to an end when sexual maturity is reached, but this may occur somewhat prematurely and abruptly if the mother conceives and gives birth again in the meantime.

A characteristic of primates, and indeed of all mammals, is that mothers feed their young on milk in their early stages of development. Here a mother vervet monkey suckles a youngster which is several months old.

During her lifetime of thirty to forty years, a female orang-utan will probably only give birth to four or five young. The young ape is not weaned for three or four years and will remain with its mother until she gives birth again.

Olive baboons generally give birth to single offspring. The young is cared for by the mother and often carried around for several months before it begins to learn to fend for itself.

The silvered leaf monkey is quite widespread in south-east Asia. The young are born with striking orange fur, which lasts for three months. After this time, the fur becomes greyish and similar to that of the adult.

Tiny and helpless at birth, this baby mountain gorilla is just starting to explore the world around it. The mother always keeps a watchful eye out, however, and usually discourages other members of the group from showing too much interest.

Baby baboons are seldom allowed to stray far from their watchful mother's gaze. This one has climbed a thorn bush in order to view the terrain and catch the first rays of the morning sun.

CONSERVATION

Although revered in some cultures and countries, monkeys and apes are under threat in most parts of the world. Threats to their survival may be direct, where animals are shot for food or artifacts, or collected from the wild for the pet trade. More insidiously and perhaps of greater importance is the threat of loss of habitat: without an environment in which to live, there is no prospect of survival for any animal.

In the rain forests of Brazil primates such as howler monkeys are killed on a regular basis for food. Indigenous people hunting by traditional means probably have little impact on numbers. However, the vast increase in human population in some areas of the Amazon, for example, and the ease with which guns can be obtained, means that monkeys generally flourish only well away from human habitation.

Regrettably, there is also a huge trade in primates as pets, both within their countries of origin and as exports. Some pet monkeys and apes are bred in captivity for this purpose. However, the very fact that a trade exists allows thousands of wild-caught specimens to be passed off under the guise of having been bred in captivity. Monkeys and apes seldom thrive in captivity: they soon show signs of depression and boredom and invariably have shortened lifespans. Nothing short of a revolution in people's attitudes toward pets and a complete ban on the trade in pet primates can solve the problem.

A greater threat to the survival of primates probably comes from habitat loss. Almost all monkeys and apes are found in the tropics, the majority occurring in rain forests or tropical deciduous forests. Logging and clearance

The establishment of forest reserves, and reintroduction to the wild of captive-bred animals, seem the only hopes for the long-term survival of the golden lion tamarin. Habitat destruction and the pet trade are the major threats to its survival.

Better known for the sound they make than their appearance, male howler monkeys advertise their group's presence at dawn and dusk with loud, growling calls. This mantled howler comes from Central America and northwestern South America.

The endearing appearance of the squirrel monkey has, unfortunately, made it popular in the pet trade. In the wild, this monkey is found from the rain forests of Brazil northward to Costa Rica.

Grooming is an uninhibited pastime performed by all primates including these patas monkeys. No region of the body is neglected and the behaviour serves to strengthen the bonds between members of the group.

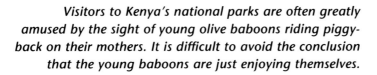

Visitors to Kenya's national parks are often greatly amused by the sight of young olive baboons riding piggyback on their mothers. It is difficult to avoid the conclusion that the young baboons are just enjoying themselves.

for agriculture are estimated to be destroying more than 50 hectares (over 100 acres) of these habitats each minute and it is not surprising that many primate species are seriously endangered.

Where the Wild Ones Are

Fortunately, the news about conservation of apes and monkeys is not all doom and gloom. In many countries, national parks and nature reserves protect the animal's environment. In some areas, the primates are protected by armed guards.

By their very nature, many of these pristine areas of forest are difficult to get to and thus it is difficult to observe their resident monkeys and apes. Some, however, offer good opportunities for observation and a few have even integrated wildlife tourism into their management plans. The following are a few of the best areas for observing monkeys and apes:

Costa Rica: white-faced capuchins are easily seen in Manuel Antonio National Park, while howler monkeys can be heard and seen at Monteverde Cloud Forest Preserve.

Brazil: bicolor tamarins and squirrel monkeys can be seen in pockets of rain forest in and around Manaus in the Amazon. Red uakaris and red howler monkeys occur around Tefe in the Amazon.

Rwanda: gorilla watching has been perfected at the Volcanoes National Park.

Tanzania: chimpanzees can be seen at the Gombe Stream reserve where Jane Goodall carried out her studies.

Malaysia: orang-utans occur at the Sepilok Forest Reserve on Sabah.

Kenya: olive baboons, vervet monkeys, and blue monkeys can all be seen within a short distance of the capital, Nairobi.

The Gambia: red patas monkeys and Guinea Forest red colobus monkeys can be approached and watched closely in the Abuko nature reserve.

Nepal: Hanuman langurs can be seen on treks from the capital, Katmandu.

As with other species of apes, grooming helps diffuse any tensions that develop between members of a group. It also allows juveniles to find their position within the hierarchy of the group.

Despite their immense size—a fully grown male may weigh 160 kilogrammes (352 pounds) or more—gorillas are in reality gentle giants. Knowing this, however, does not dispel a feeling of intimidation during a face-to-face encounter.

INDEX

*Page numbers in **bold-face** type indicate photo captions.*